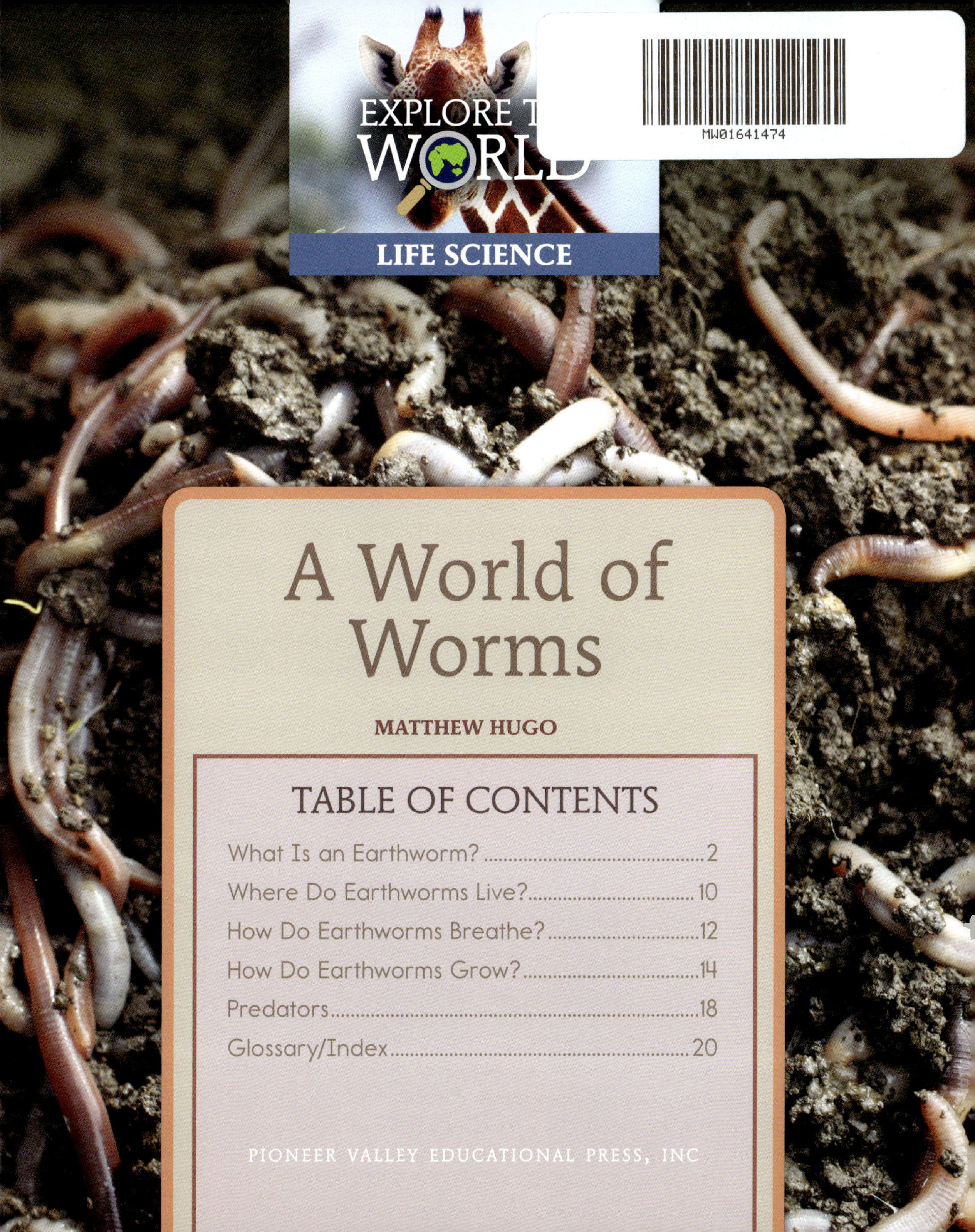

A World of Worms

MATTHEW HUGO

TABLE OF CONTENTS

PIONEER VALLEY EDUCATIONAL PRESS, INC

WHAT IS AN EARTHWORM?

If you dig a hole in the dirt,
there is a good chance
you will find something slimy and wriggly.
It's an earthworm!
Earthworms are worms that live in the **soil**.

MORE TO EXPLORE

An earthworm is sometimes called a **NIGHT CRAWLER**.
What time of day do you think you would see a night crawler?

Earthworms do not have arms, legs, or eyes.
They are soft and long.
Most earthworms are reddish brown.
They have over 150 parts called **segments**.
On each segment are little hairs.
The hairs help the earthworm move.

MORE TO EXPLORE

Earthworms can be as small as half an inch or as long as 10 feet!

Earthworms make tunnels
through the soil
called **burrows**.
The burrows help the soil get air and water.
Earthworms eat plants and soil
as they move along.
The plants and soil
pass through the earthworm.
The earthworm leaves behind piles
of the eaten soil
and plants called **castings**.
These castings make good plant food
and help fruits and vegetables to grow.

An earthworm can eat up to one third of its size in a day.

At night, an earthworm will stretch its body out of the dirt and pull food back into its burrow. It keeps its back end in the burrow so it can disappear quickly if a **predator** tries to eat it.

WHERE DO EARTHWORMS LIVE?

You can find earthworms in most places around the world. They live in fields, woods, and gardens. They do not live in deserts or in places where there is ice and snow all year. Most earthworms live underground.

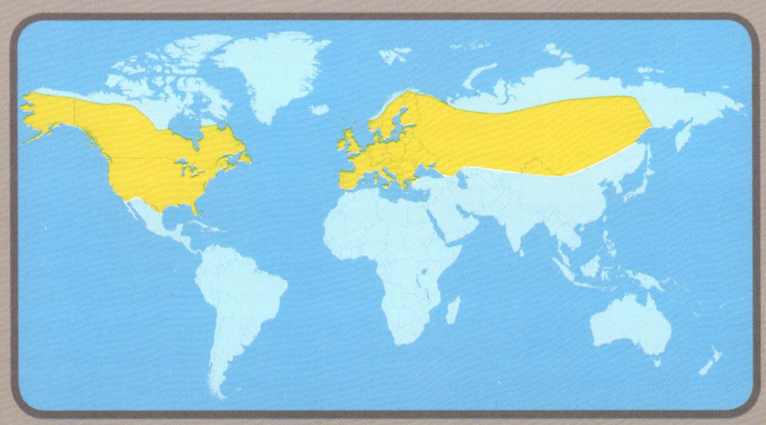

There are more than 6,000 different kinds of earthworms.

HOW DO EARTHWORMS BREATHE?

An earthworm does not have lungs.

It breathes through its skin.

An earthworm needs to have **damp** skin to breathe.

If a worm's skin dries out, it will die.

Earthworms usually come out at night when the air is cooler and they will not dry out.

HOW DO EARTHWORMS GROW?

After an earthworm lays an egg,
a cocoon forms around
the egg to protect it.
The cocoon is buried in the dirt.

The baby earthworms will hatch
in two or three weeks.
The baby worms are white
and very small.
They will grow quickly
and turn into adult worms
in just four to six weeks.

A hungry bird ate this earthworm's tail,
but the worm can grow a new one.
The part with the head
can grow new parts again.

PREDATORS

Earthworms have many predators.
There are lots of animals
that like to eat earthworms.
Birds, rats, and toads like to eat earthworms.
Snakes, foxes, and bears
will also eat earthworms.

bird
mole
toad
turtle
fox
bear

GLOSSARY

burrows

holes or tunnels

castings

piles of plants and soil left behind by earthworms

damp

a little wet

predator

an animal that lives by killing and eating other animals

segments

parts cut off from

soil

the loose dirt where plants grow

INDEX

How to Make a Worm Farm

What you need:

- 2-liter bottle
- scissors (parent cuts!)
- tape
- small rocks
- sand
- dirt
- leaves
- worms
- black paper

1 A parent cuts the top off the bottle.

2 Fill as described below:

2 inches of dirt with some leaves and worms

2 inches of sand

2 inches of dirt with some leaves and worms

2 inches of sand

2 inches of rocks

3 Tape the top back on and cover the bottle with black paper. (Worms like the dark.) Peek in on them to see their activity.

GLOSSARY

burrows
holes or tunnels

castings
piles of plants and soil left behind by earthworms

damp
a little wet

predator
an animal that lives by killing and eating other animals

segments
parts cut off from

soil
the loose dirt where plants grow

INDEX